STEM CAREERS

ANIMAL PLANT and VETERINARY Sciences

C. W. Wilson

STEM CAREERS

ANIMAL PLANT and VETERINARY Sciences

ISBN 978 1 304 93436 9

Copyright © 2013, 2014 by C. W. Wilson
Cover Design – C. W. Wilson
Formerly STEM CAREERS Working with Animals and Plants

CONTENTS

What are Science STEM Careers?

S.T.E.M. is an acronym for disciplines and careers in the **Sciences, Technology, Engineering and Mathematics**. Those in the **STEM Science Careers** who work with **animals and plants** are at the forefront of preserving resources and are researching future food and water sources. As one of these scientists discoveries can play a major role in maintaining and sustaining earth's natural resources.

Scientists who work with animals and plants within some Careers are on the leading edge of preserving earth's natural resources. This book allows anyone interested in these animal and plant science careers to make informed career decisions.

- ❖ **CAREER DESCRIPTIONS**, **EDUCATIONAL** and **ACADEMIC REQUIREMENTS**, **JOB OUTLOOK** and long-term career information is provided for those interested in working with Animals and Plants.

Included as a BONUS!

- ❖ A **Bonus** – Included are **Easy to Follow Tools** helping you to discover the science careers matching your natural cognitive skills. These tools are the **COGNITIVE INDEX©2013** and **SKILLS SET INDEX©2013** and also the **WORKING ENVIRONMENT GUIDE©2013.**

STEM Scientists' work impacts plant and animal life in meaningful ways and has positive long-term effects for our planet.

The Author

SCIENCE CAREERS by DISCIPLINES

Agriculture and Animal Scientists

**AGRIBUSINESS   AGRONOMY
ANIMAL BREEDERS**

Botanists

**BIOCHEMISTRY   BIOPHYSICS
BIOTECHNOLOGY   CYTOLOGY
ECOLOGY   FOOD TECHNOLOGY   FORESTRY
GENETICS   HORTICULTURE
MOLECULAR BIOLOGY   PALEOBOTANY
PHYSIOLOGY   PHYTOCHEMISTRY
PLANT PATHOLOGY   TAXONOMY**

Conservation Scientists and Foresters

**BIOCHEMISTRY   GENOMICS   HYDROLOGY
MOLECULAR BIOLOGY   MOLECULAR GENETICS
ORGANIC CHEMISTRY
RANGE CONSERVATIONISTS
RANGE ECOLOGISTS   RANGE MANAGERS
RANGE SCIENTISTS   SOIL CONSERVATIONISTS
QUANTUM BIOLOGY
WATER CONSERVATIONISTS**

Environmental Scientists

**ENVIRONMENTAL CHEMISTS
ENVIRONMENTAL HEALTH SPECIALISTS
ENVIRONMENTAL PROTECTION SPECIALISTS**

SCIENCE CAREERS by DISCIPLINES
(continued)

Food Scientists

**FOOD CHEMISTRY   FOOD DESIGN
FOOD ENGINEERING   FOOD MICROBIOLOGY
FOOD PACKAGING   FOOD PHYSICIST
FOOD PRESERVATION
FOOD PRODUCT DEVELOPMENT
FOOD SAFETY SENSORY ANALYSIS**

Veterinarians

**COMPANION ANIMAL VETERINARIANS
CLINICAL VETERINARIANS
EQUINE VETERINARIANS
FOOD AND SAFETY VETERINARIANS
FOOD ANIMAL VETERINARIANS
RESEARCH VETERINARIANS**

Wildlife Biologists and Zoologists

**BIOCHEMISTS ECOLOGISTS
ENTOMOLOGISTS
EVOLUTIONARY BIOLOGISTS
HERPETOLOGISTS   LIMNOLOGISTS
MAMMALOGISTS   MARINE BIOLOGISTS
MICROBIOLOGISTS   ORNITHOLOGISTS
PHYSIOLOGISTS   WILDLIFE EDUCATORS
WILDLIFE RESEARCHERS
WILDLIFE REHABILITATORS**

SUCCESSFUL

SKILLS

FOR

SCIENTISTS

Table 1

COGNITIVE INDEX© 2013

Everyone has a predominant Cognitive Style.   A Cognitive Style includes instinctive thinking which is natural to an individual. Instinctive Cognitive Styles are many and can include Practical, Logical, Independent, Creative and Artistic styles along with other styles.

Very successful scientists in all concentrations exhibit Practical and Logical cognitive styles and have the ability to creatively interpret and apply data and information. Making use of the following information to match your natural cognitive skills with those valued in your work environment can assist in attaining a successful career.

PRACTICAL	Applying information and knowledge towards tasks in a manner that is useful and beneficial.
LOGICAL	Considering whether data, information and facts are valid, invalid or hold a portion of validity in efforts to reach known and reproducible results.

COGNITIVE INDEX (continued)

INDEPENDENT
Working alone to research or produce a project or projects. In some cases natural instinct could be to test and reject traditional or current methods of performing working activities.

CREATIVE
Originating, imaging or producing solutions, products, works or items of value in a unique and original manner.

ARTISTIC
Exhibiting and applying insight and knowledge while using creative skills in physical execution of an item or thing.

Table 2

SKILLS SET INDEX© 2013

Those who are successful in their chosen careers no matter the area of concentration apply their natural interests and abilities within their chosen career. An example of this can be Veterinarians who must perform surgery on animals have natural skills such as excellent critical thinking, analytical ability. They also have very good dexterity ability to perform this work.

Scientists also have the ability to master other necessary skills such as becoming an effective team player. These can be considered "soft skills" or "people skills. Other soft skills can include creative abilities to view research data and information from different perspectives along with collaborating and sharing observations and ideas which can provide new insights and discoveries. The following lists some qualities and necessary attributes assisting successful scientists.

SUCCESSFUL SKILLS FOR SCIENTISTS

RESEARCH and LOGICAL SKILLS

- **CRITICAL THINKING** skills include data-analysis abilities, successfully using both qualitative and quantitative research and study methodologies.

SKILLS SET INDEX (continued)

- **ADVANCED MATHEMATICAL** skills include understanding and executing complicated and complex algebra, geometry and calculus calculations. Much of their research requires expression in mathematical terms.

- In depth **ANALYTICAL SKILLS** requires precision and accuracy when researching, providing data and validating research findings.

- Excellent **DECISION MAKING** skills are necessary to understand information and to accurately interpret findings from data analysis.

- **OBERVATION SKILLS** are necessary to decisively connect research information and data to findings and results.

- **CREATIVELY SOLVING COMPLEX PROBLEMS** including abilities to successfully design, plan and implement projects.

- **WRITING SKILLS** include an ability to write findings clearly and concisely in well organized and clearly written reports.

SUCCESSFUL SKILLS FOR SCIENTISTS

SOFT SKILLS

- **EXCELLENT COMMUNICATION SKILLS** are necessary to explain results from studies, methodologies and scientific discoveries to effectively explain implications from findings to peers, corporate members and in some cases the general public.

- Effective **TEAM PLAYERS** posses and practice vital skills including abilities to work with and share ideas and data with others including peers, clients and management.

About Working Environments

Every workplace whether service oriented or manufacturing has a specific political or working environment. **(See Table 3)** The nature of "political environments" or "office politics" in the workplace is rarely a topic open for discussion.   "Office politics," "political environments" and "working environments" are used interchangeable and have similar meanings.

Humans are very political by nature as we socialize with friends and family; enjoy recreational activities and also when we work.

Political in this sense does not have the same meaning as a person running for a political office but the means and methods a person in the workplace typically interacts with another. Anyone who has worked for a number of years could be aware, whether consciously or not, his or her place of work has a particular "political environment" or working environment. Those new to the workplace can be surprised when discovering the level of "office politics" occurring in the workplace.

Those who are currently successful and excelling in a particular working environment are typically in tune with their current workplace "political" environment. Many have learned to successfully adapt to the working or political environment leading to their career success.

About Working Environments (continued)

An example could be you enjoy challenging authority as you discover innovative ways of reaching your goals but the "working environment" of your chosen organization only rewards those who exhibit patience and a low key personality. This difference between the attributes the organization considers to be valuable and your thinking and cognitive style could lead to frustration where your work could eventually suffer within an incompatible working environment.

It is beneficial to have an idea of the variety of "political" and workplace environments within various organizations as you research the career of your choice. This information is provided as a very general guide and it is your choice to make use of this guide.   It is important for you to make your own decisions while making informed choices regarding your career.   (See **WORKING ENVIRONMENTS GUIDE**)

Table 3

WORKING ENVIRONMENT GUIDE© 2013

Most workplaces are a mixture of one or more environments. Listed here are some definitions of usual working and "political" environments:

HIGHLY POLITICAL — HP
Perceived power and ambition is highly valued.

MODERATELY POLITICAL — MP
Perceived power or ambition may or may not be rewarded.

LEAST POLITICAL — LP
Ambition is not an asset or a liability. This can be defined as those who are viewed as ambitious with an aim to get ahead are judged similarly as someone who does not appear to be ambitious.

BUREAUCRACY/CONFORMITY   VALUED — BUR/CF Following working procedures which are valued by an organization's management is usually rewarded.

NEW IDEAS WELCOMED — NI
Managers, executives and top decision-makers value and reward new ideas. Small or newly formed private working environments can be more open to new ideas **(NI)**.

STARTUP ORGANIZATIONS — SU
Independence and new ideas are welcomed and rewarded.

Table 3.1.

WORKING ENVIRONMENTS TABLE

This is a general list of some environments where scientists work.   Refer to **Table 3** for definitions of Working Environments.

CAREER TYPE   ENVIRONMENT

Colleges and Universities	BUR/CF
Consulting Firms	HP, BUR/CF, NI*
Corporate Laboratories	HP, BUR/CF
Educational Laboratories	BUR/CF, MP
Environmental Companies	BUR/CF
Farms	BUR/CF, HP
Government Organizations	BUR/CF, HP
Hospitals	HP, BUR/CF/CF
Manufacturing Companies	BUR/CF/CF, HP
Parks and Zoos	BUR/CF, HP
Private Engineering Companies	HP, BUR/CF
Private Research Institutions	HP, BUR/CF
Processing Companies	HP, BUR/CF
Research Organizations	HP, BUR/CF
Veterinarians Offices	HP, NI *

*These are in small or newly formed private working environments which can be more open to new ideas and independence among peers and management within these environments.

SCIENCE

STEM

CAREERS

Highlighted SCIENCE CAREERS

❖ If you have a strong interest in a career dedicated to understanding food quality and food processes **Agricultural** and **Food Scientists** careers could be for you.

❖ A **Food Scientist** can also be a **Food Regulator** working within the agricultural discipline and within government organizations enforcing food regulations for restaurants and commercial organizations.

❖ A focus on the functions, structures and growth of plants could lead you to consider becoming a **Botanist** as a career choice.

❖ If understanding the air, soil and water is of interest work in the area of **Ecology as an Environmental Scientist** is one discipline to choose. Other work could involve assisting in chemical spills and understanding how pollutants negatively impact land and water quality.

❖ If you enjoy preserving natural resources, forests and land you could be a **Conservation Scientist** or **Forester** who works very closely with others with academic concentrations in ecology.

❖ **Veterinarians'** goals are to provide care along with maintaining and improving the health of pets, horses, livestock and animals in zoos. Practicing Veterinarians' education and training is just as intense as most scientists.

❖ **Wildlife Biologists and Zoologists** work to research, understand, improve and preserve quality of life for wildlife when possible. These scientists research the health, behavior and activities of mammals, birds, fish, reptiles, amphibians and insects within natural habitats and in controlled environments such as preservation parks and zoos.

AGRICULTURAL SCIENTISTS

What do they do?

Agricultural Scientists use knowledge of biology, chemistry, physics, mathematics, and other sciences to provide agricultural solutions. Some **Agricultural Scientists** research and discover the most current methods to raise and breed farm animals while increasing productivity of farm animal products. These products include meat, milk and egg production. **Agricultural Scientist** should be very interested in working with farm animals, crop production techniques with a desire to attain and maintain the highest quality within their work. Animal Scientists and Agricultural Scientists duties are similar as they research, analyze and monitor farm animals' biology, nutrition and disease control methods.

Agricultural and Food Scientists work to understand and improve the biological and chemical makeup of food which is consumed along with new food products. They also work with crop production and soil management which is **Agronomy**, a branch of agriculture dealing with the development and application of high-quality food, crop production and soil management.

Others work with **Agribusinesses** having goals of planting, harvesting and researching agricultural for human consumption.

Some Agricultural Scientists conduct inspections ensuring humane and sanitary facilities are available to farm animals. Agricultural Scientists can also educate farmers and ranchers on methods to improve conditions when necessary. In many cases Agricultural Scientists work closely with Biological Scientists applying research techniques and methods.

Research methods include testing and monitoring feed and nutritional quality such as milk production from cows; quality of water and soil pollutant levels; along with monitoring the quality and quantity of animal's reproductive levels. Agricultural Animal Scientists are concerned with farm animals' quality of life including disease and pest control.

Where Do Agricultural Scientists Work?

Many Agricultural Scientists work in research and development within private and academic laboratories. Some Agricultural Scientists work in the field providing hands-on research and others work for corporate organizations. A good amount of time is spent analyzing, studying and documenting data and reports. High tolerance of a variety of environments is necessary due to regular visits to farms and food manufacturing plants.

Agricultural and Food Scientists perform very similar duties and activities allowing the work performed in these careers to become interchangeable. Other areas Agricultural and Food Scientists work within involve using current technology to preserve and process food.

Many with bachelor's degrees in agricultural sciences find work in related jobs instead of becoming agricultural scientists. For example, a bachelor's degree in agricultural science is useful for managerial jobs in farm-related or ranch-related businesses such as farming, ranching, agricultural inspection, farm credit institutions, or companies making or selling feed, fertilizer and farm equipment.

Employment of Agricultural and Food Scientist is expected to increase. This field does not employ large numbers of people. About 3,000 people are currently Agricultural and Food Scientists in the United States. Due to the relatively low number of Animal, Agricultural and Food Scientists normally employed in this field the hiring potential and landing a position in this field is **Highly Competitive.**

What type of education does it take to become an Agricultural Scientist?

Look for colleges and universities offering degrees in agricultural science. In the United States most states have a land-grant college offering animal science, agriculture and food sciences curriculum. A bachelor's degree in agricultural science can lead to work in farming, agricultural inspection or in companies manufacturing farming equipment. Undergraduate courses in biology, physics, engineering and chemistry can also lead to careers as Agricultural Scientists. Some positions including product developers require a bachelor's degree in agricultural science.

Advanced research requires a Ph.D. in a related field of study. In order to compete for the limited number of available Animal Scientists positions it is best to attain a Ph.D. which can take six years to complete. Some earn a Doctorate of Veterinary Medicine (DVM) for greater chances for success. In order to teach at universities and colleges a masters degree or Ph.D. is needed.

Graduate school provides advanced coursework including laboratory work and original research experience.   Animal reproduction, biotechnology, genetics, statistical analysis and experimental design are subjects offered at the Ph.D. level.

When you graduate how many jobs may be available as an Agricultural Scientist?

Employment of Agricultural and Food Scientists is expected to increase by ten percent (10%) through 2020. A high interest in food safety is fueling an increase in the need for Agricultural and Food Scientists.

Most growth over the next ten years (10) for Animal, Agricultural and Food Scientists will be in private industry. Private industry has increased its demand for agricultural and food scientists due to a demand for more in-depth knowledge and data regarding crop development, developing food and pharmaceuticals along with demands for greater food quality and safety.

The average salary for Agricultural and Food Scientists for state, local and private colleges and universities after some experience in the field can be about $46,000. With a Ph.D. in these areas those working within private organizations can earn annual salaries of $57,000 or more.

With an advanced degree such as a doctorate (Ph.D.) in this field work could also involve genetic research and application.

Understanding ecosystems and how pest infiltrate and impact both water and soil is leading to higher demands for these scientists. Some research by these scientists can be performed using technology and biotechnology. Biofuels derived from plants and other renewal energy sources are also driving a need for more Agricultural Scientists.   Interest in and demand for sustainable practices such as using animal waste instead of chemicals to naturally sustain environments has also increased.

How can your Cognitive thinking Style and Skill Set assist in determining your success in working environments as an Agricultural Scientist?

Successful Agricultural Scientists have abilities and Skill Sets **(Table 2)** such as excellent **critical-thinking, analytical** and **observation skills** and **advanced mathematical ability.**   They are able to creatively interpret and apply data and information into documents and projects while using excellent report writing skills. Having **Practical** and **Logical (Table 1)** Cognitive styles can be very beneficial as an Agricultural Scientist. These scientists explain their findings including the methods used and implications of the results.   They have also mastered the "soft skills" of expressing good communication skills and are usually a productive team player.

Working Environments **(Table 3)** in this career can range from **Highly Political (HP)** to **Least Political (LP).**   In many academic and government positions working environments can be both **Highly Political** and **Bureaucracy/Conformity (BUR/CF)** can be valued and rewarded. Manufacturing facilities can also be **Highly Political** and **Bureaucracy (BUR/CF)** is usually valued.   Research and Development areas within these organizations could be more open to **New Ideas (NI).**

Private research and development organizations could also be very **Highly Political** and in some organizations new methods are rewarded.   Small or newly formed private working environments can be more open to new ideas **(NI).**

Where Agricultural Scientists Work

Organizations and working environments vary regarding degree requirements. The information below is provided for general information only. Look into your area of interest for current degree requirements.

SPECIALTY   DEGREE NEEDED

Farming
Farm Managers   Bachelors and Higher
Feed Specialist   Bachelors and Higher
Pest Control Chemists   Bachelors and Higher

Agricultural/Research
Agricultural Scientists   Bachelors and Higher
Food Scientists   Bachelors and Higher
Government Regulators   Masters and Higher
Product Developers   Bachelors and Higher

Academic Institutions
Academic Instructors   Masters or Doctorate
Academic Researchers   Doctorate

Manufacturing
Agricultural Inspector   Masters and Higher
Animal Breeder   Bachelors and Higher

Ranching
Artificial Insemination
Specialists   Masters and Higher
Breeding Technicians   Bachelors and Higher

Forestry
Conservation Scientists   Bachelors and Higher
Land Managers   Bachelors and Higher
Procurement Foresters   Bachelors and Higher

ANIMAL SCIENTISTS

What do they do?

Animal Scientist's work is performed within many environments. **Animal Scientists** research nutrition, genetics, reproduction and development of domestic farm animals. Eggs, poultry, meat and milk product processes are studied and researched assisting in greater safety and better disease control methods increasing animal production and products.

Animal Scientists use research methods while analyzing and documenting findings. Some also breed selected animals assisting to attain the most desirable traits and characteristics of certain animals for production. Investigation and analysis include processing and feeding methods and techniques. Analyzing environmental conditions and its effect on the quality of animal and animal products is also researched, analyzed and documented. Work in this area includes researching nutrition, disease control methods, reproduction of eggs from chickens and milk from cows and a variety of animals for the production of food and food products.

Major differences between Agricultural and Animal Scientists and Zoologists are their working environments can differ. Agricultural and Animal Scientists work can involve researching pest and disease control for animals raised for food production while Zoologists work can involve emphasis on animal care and preservation within controlled environments, zoos and in natural habitats.

Where Do Animal Scientists Work?

Animal Scientists work in businesses, government organizations, education, and research facilities.

Private companies and governmental organizations employ the most Animal Scientists.   Almost half of Animal and Agricultural scientists work for government, public universities and private colleges performing development and research work.   Agricultural and Animal Scientists working in some private organizations including food manufacturers and food processing plants generally research the best methods to breed and raise farm animals. These scientists' goals include increasing productivity of animal products for consumption and other uses including wool and leather.

If an Animal Scientist works with cows for milk production his or her goal might be to discover the most effective types of nutrition, living conditions and care to achieve the highest quality milk production. Scientists working with chickens apply similar techniques for the highest quality and greatest quantity of eggs and chicken for production.

Many with bachelor's degrees in animal and agricultural sciences find work within related jobs instead of becoming an Animal or Agricultural Scientist. For example, a bachelor's degree in agricultural and animal science is useful for managerial jobs in farm-related or ranch-related businesses, such as agricultural inspection, farming, farm credit institutions, ranching or companies making or selling farm equipment, feed, and fertilizer.

What type of education does it take to become an Animal Scientist?

Look for colleges and universities offering degrees in agricultural science. In the U.S. most states have a land-grant college offering animal science, agriculture and food sciences curriculum. A bachelor's degree in agricultural science can lead to work in farming, agricultural inspection or in companies manufacturing farming equipment. Undergraduate courses in biology, physics, engineering and chemistry can also lead to careers as an Agricultural Scientist after degrees are earned. Some positions including product development require a bachelor's degree in agricultural science.

In order to compete for the limited number of available Animal Scientists positions it is best to attain a Ph.D. which can take six years to complete. Advanced research requires a Ph.D. in a related field of study. Some earn a Doctorate of Veterinary Medicine (DVM) for greater chances of success. To teach at universities and colleges a master's degree or Ph.D. is required.

Graduate school provides advanced coursework including laboratory work and original research experience. Animal reproduction, biotechnology, genetics, statistical analysis and experimental design are subjects offered at the Ph.D. level of study.

When you graduate how many jobs may be available as an Animal Scientist?

Animal Scientists, Agricultural and Food Scientists positions combined are expected to increase about ten percent (10%) through 2020. The demand for healthier food and more ecosystem-friendly methods of attaining food products is driving the need for more research in these areas.

Most growth over the next ten (10) years for Animal, Agricultural and Food Scientists will be in private industry. Private industry has increased its demand for agricultural and food scientists due to a demand for greater expertise in developing food, crops, and pharmaceuticals, along with ensuring food quality and safety. In the United States after earning academic degrees certifications recognizing expertise in this field are available. **(See Information and Resources)**

After degrees are attained Animal Scientist salaries can start around $32,000 annually. The average salary after some experience in the field can be about $58,000. The highest salaries in this area are around $117,000 annually.

Employment of Animal and Agricultural Scientists is expected to increase. This field does not employ large numbers of scientists. About 3,000 people are currently Animal, Agricultural and Food Scientists. Due to the relatively low number of Agricultural and Food Scientists normally employed in this field the hiring potential and landing a position in this field is **Highly Competitive**.

Greater interests in understanding ecosystems and how pest infiltrate and impact plants, water and soil are leading to higher demands for these scientists. A variety of information regarding how to maintain safe and healthy feed for production animals are also driving a need for more Animal Scientists. Interest in and demand for sustainable practices such as using animal waste instead of chemicals to naturally sustain environments has also increased.

These scientists should also be very comfortable around animals as their work could occur in confined areas. Greater demands for food safety and interest in developing new pharmaceuticals increase the need for more Animal Scientists.

How can your Cognitive thinking Style and Skill Set assist in determining your success in working environments as an Animal Scientist?

Successful Animal Scientists have abilities and Skill Sets **(Table 2)** such as excellent critical-thinking, analytical and observation skills and advanced mathematical ability along with the ability to creatively interpret and apply data and information into projects, documentation, presentations and reports. Very good communication skills are also highly valued. These scientists explain their findings including the methods used and implications of the results. They have also mastered the "soft skills" of being a productive team player. Having **Practical** and **Logical (Table 1)** Cognitive thinking styles can be very beneficial as an Animal Scientist.

Working Environments **(Table 3)** in this career can range from **Highly Political (HP)** to **Least Political (LP).** In many academic and government positions working environments can be both **Highly Political** and **Bureaucracy/Conformity (BUR/CF)** can be valued and rewarded. Manufacturing facilities can also be **Highly Political** and **Bureaucracy (BUR/CF)** is usually valued. Research and Development areas within these organizations could be more open to **New Ideas (NI).**

Private research and development organizations could also be very **Highly Political** and in some organizations new methods are rewarded. Small or newly formed private working environments can be more open to new ideas **(NI).**

Animal Scientists

Areas of Concentration

Organizations vary regarding necessary degree requirements. Work experience in specific areas can impact the type of degree required.   Look into your area of interest for degree information. Data below is provided for general information only.

SPECIALTY   DEGREE NEEDED

Farming and Ranching
Farm Managers   Bachelors and Higher
Feed Specialist   Bachelors and Higher
Pest Control Chemists   Bachelors and Higher
Pest Control Specialist   Associates and Higher
Breeding Technicians   Bachelors and Higher

**Agricultural Inspection
and Research**
Agricultural Scientists   Bachelors and Higher
Food scientists   Bachelors and Higher
Government Regulators   Masters and Higher
Product Developers   Bachelors and Higher

Academic Institutions
Academic Instructors   Masters or Doctorate (PH.D.)
Academic Researchers   Doctorate (PH.D.)

Manufacturing
Agricultural Inspector   Masters and Higher
Animal and Livestock Breeder   Bachelors and Higher
Feed and Equipment   Bachelors and Higher
Pharmaceutical Organizations   Masters and Higher
Meat Processing Companies   Bachelors and Higher
Food Distributing Companies   Bachelors and Higher
Magazines and Writers   Bachelors and Higher

BOTANISTS

What do they do?

Botanists study plants to improve quality of life for plants, animals and humans. Botany is the scientific study of plants and plant organisms from seeds to giant trees. Research by these scientists improves pharmaceutical medicines along with food and food products. These scientists also research natural fibers and building materials from trees and plants. Botanists' work falls in many areas including Conservation, Environmental Protection and Public Health.

Some **Botanists** are also called **Plant Biologists** and they work within a wide area of concentration. Some **Ecology Botanists** concentrate on interactions of plants with their environment including animals and humans. Many Botanists perform fieldwork searching and discovering new species of plants. Other Botanists or Plant Biologists research and analyze the structure of plants. Many Botanists perform experiments to discover minute details of plants and how plants convert from simple compounds to complex chemicals.

Botanists also study how DNA works in plants and how evolutionary genetic information is created and developed as they study plants' evolution over large amounts of time.

A number of Botanists work with Conservation Scientists assisting with park management. They analyze and share their findings with those impacted by their decisions. Foresters can also use information acquired by Botanists to maintain forests, rangelands and wilderness areas. Botanists' research can be very helpful for public health agencies and Environmental Protection workers assisting to solve pollution and contamination problems.

Botanists along with Conservation Scientist, Environmental Scientists and Foresters work with a 'big picture" viewpoint of a variety of factors affecting natural environments.   These factors include an ecological balance between human expansion and activities and natural resources including plants, trees, land, animals and soil preservation within these environments.

Where Do Botanists Work?

Some Botanists work with one group of plants. **Agronomists** may research and analyze agricultural crops. **Marine Botanists** research plants growing in the ocean.   **Plant Botanist** and **Physiologists** concentrate on the full life processes and cycles of a variety of plant species. **Economic Botanists** research and develop plants used for commerce and can be sold as pharmaceuticals, food and other products.

Most Botanists work within college and university environments.   In these environments they conduct research and teach.   Government organizations also employ many Botanists within research and development and within laboratories. A smaller number of Botanists work for private organizations including botanical gardens.   Work can involve performing fieldwork within specific areas.   Some Botanist work alone and others work within a research team. Some of these scientists travel to sites collecting plant specimens.

Botanists should be highly curious about plants and plant life cycles, structures and development. Most Botanists work and exchange research data and findings with other scientists. Having good communication and writing skills are highly valued in this career.

These scientists also work in and within a variety of specialties including **Soil** and **Water** and **Forest Protection** and can involve a good amount of fieldwork in areas such as prairies and woodlands; many work in remote and hard to reach areas.

What type of degree do you need to become a Botanist?

Many Botanists major in biology and earn a bachelor's degree. With a bachelors degree Botanists can work as technical writers or biological technicians among a variety of careers.

Some specialize in botany and plant physiology while earning a masters or doctorate (Ph.D.). With a masters degree you can begin a career within entry-level research positions. These positions may be limited with either a bachelor's or master's degrees.

For greater advancement in this field and for more lucrative career opportunities those most successful in this field have a doctorate degree (Ph.D.). A Ph.D. is required to teach Botany at most colleges and universities. In this area many new discoveries often make it necessary to study throughout a Botanist's entire career.

After years of research experience some can become directors of research at government organizations and within private organizations. Those teaching at colleges and universities can become professors. Some publish their research findings in plant and botanical related journals.

Research your options to ensure the school you are attending or plan to attend is an accredited school in the field of your study and degree.

When you graduate how many jobs may be available as a Botanist?

Employment of Botanists is expected to increase about eight percent (8%) through 2020. Due to the small amount of new Botanists positions which could be available this is a **Highly Competitive** career for many reasons including recent budget and funding cuts by the government. Teaching positions at colleges and universities are also **Highly Competitive**.

Intern positions and work-study opportunities could be available within colleges and universities research laboratories, government organizations, agricultural and biological laboratories and within private organizations.

Salaries for Botanists can range from around $41,000 for entry-level positions to $89,000 or more for experienced Botanists. For those with the required degrees and more experience working for colleges, universities and research laboratories salaries could reach around $98,000 annually.

Interest in understanding our ecosystems along with balancing human activities and preserving natural environments is leading the way for a slight increase in demand for **Botanists, Environmental Scientists, Foresters** and **Conservation Scientists**.

How can your Cognitive thinking Style and Skill Sets assist in determining your success in working environments as a Botanist?

Successful Botanists have many practical abilities along with excellent underline critical-thinking, analytical and advanced mathematical ability. They are able to creatively interpret and apply data and information into projects, documentation and reports.

Successful Botanists have abilities and Skill Sets **(Table 2)** such as excellent critical-thinking, analytical and observation skills and advanced mathematical ability. Very good communication skills are also highly valued.

Interest in nature preservation along with keen observation skills to detect small changes and patterns in natural environments are some of the most important skills for successful Botanists. Great observation skills can be beneficial for Botanists as the slightest changes in plants are observed and analyzed. **Practical** and **Logical** Cognitive **(Table 1)** thinking styles can be very beneficial as a Botanist. Successful Botanists also possess the ability to creatively solve complex problems especially when determining the most effective route to make useful and applicable decisions concerning plant preservation.

Working Environments **(Table 3)** in this career can range from **Highly Political (HP)** to **Least Political (LP).** In many academic and government positions working environments can be both **Highly Political** and **Bureaucracy/Conformity (BUR/CF)** can be valued and rewarded. Manufacturing facilities can also be **Highly Political** and **Bureaucracy (BUR/CF)** is usually valued. Research and Development areas within these organizations could be more open to **New Ideas (NI).**

Private research and development organizations could also be very **Highly Political** and in some organizations new methods are rewarded.   Small or newly formed private working environments can be more open to new ideas **(NI).**

Botanists

Areas of Concentration

ANATOMY and BIOPHYSICS – Study of microscopic plant structure, processes, cells and tissue.

BIOCHEMISTRY and **PHYTOCHEMISTRY** – They study chemicals and processes within plants.

CYTOLOGY – Study of functions and history of plants at the cell level.

ECOLOGY – Study of interaction and impact of humans and animals upon plants.

GENETICS and **MOLECULAR BIOLOGY** – Study of plant cells at the RNA and DNA levels.

PALEOBOTANY – Study of history, structure, biology and evolution of fossil plants.

PHYSIOLOGY – Study of the functions of plants such as photosynthesis and how plants receive nutrition.

TAXONOMY – Study involves identifying, naming and classifying plants.

Disciplines in Plant Sciences

There a variety of careers in the area of plant research, analysis and observation. Some are below:

AGRONOMY – Those in this fieldwork to make beneficial use of plant and soil science for crop production.

BIOTECHNOLOGY – Using genes to produce new or desired results in plants.

FOOD SCIENCE AND TECHNOLOGY – Research, analyze and develop food and food products from a large array of plant products.

FORESTRY – Managing forest species for conservation purposes and timber production.

HORTICULTURE – Produce ornamental plants including vegetable and fruit crops.

NATURAL RESOURCE MANAGEMENT – Structured protection duties for the protection of natural resources.

PLANT PATHOLOGY – Study the diseases of plants including prevention and solutions.

Careers in Botany

Ecologist, Conservationist, or Forester might be for you if you enjoy being outdoors a good amount of time. There could be a lot of traveling to different locations and various parts of the country and also to different countries.

Biophysics, Developmental Botany, Genetics - If you have strong mathematical tendencies and a strong interest in plants these fields could interest you.

Microbiology or Mycology - Understanding the microscopic structures of plants is another area in field of Botany.

Ornamental Horticulture - It you have creative and artistic interests in plant forms for display this career could be for you.

Plant Biochemist, Plant Physiologist, Molecular Biologist or Chemotaxonomist - With a strong interest in plants and chemistry these careers might be of interest.

Plant Breeding - An interest in the process of plant cell nourishment processes and food supplies and plant preservation is another area a Botanist could find of interest.

Plant Pathology — These Botanists analyze and study the diseases of plants.

CONSERVATION SCIENTISTS

What do they do?

Conservation Scientists work to research, understand and implement ecologically safe relationships between nature and humans. **Conservation Scientists** research and manage natural resources while balancing human needs and activities regarding land and ownership.

Conservation Scientists work with a 'big picture" viewpoint of a variety of factors affecting natural environments. These factors include an ecological balance between human expansion and activities and natural resources including plants, trees, land, animals and soil preservation within these environments.

Conservation Scientists work closely with and can also advise crop landowners, farmers and property owners to ensure natural land preservation meets their needs. These scientists provide specific knowledge while developing conservation strategies to maintain biological diversity among plants and animals in nature. They also work with residential and commercial property owners for recreational purposes such as entertainment parks.

Where Do Conservation Scientists Work?

Conservation Scientists work for both government and private organizations. With experience some work independently. Careers in government include performing studies regarding areas impacted by human expansion and activities; other studies can include researching and recommending ecological benefits of re-forestation in negatively impacted areas.

Private organizations including private landowners may use Conservation Scientists to assist and research to discover the best use of privately owned natural areas.

These scientists work in a variety of specialties including Soil and Water and Forest Protection, and can involve a good amount of fieldwork in areas such as prairies and woodlands; many work in remote and hard to reach areas.   These scientists are also involved in decisions regarding both personal and public land use.   Some scientists within this career also work in offices.

What type of degree do you need to become a Conservation Scientist?

Research available options to ensure the school you are attending or plan to attend is **an accredited school in the field of your study and degree**. Most scientists in this field have earned a bachelor's degree along with higher degrees in forestry, rangeland management, and environmental or agricultural science or in Natural Resources or Ecology and Management.

Bachelor degrees in Environmental Science along with bachelor degrees in Natural Resources Ecology and Management are available in some areas. Colleges and universities in many states offer forestry degrees.

Masters in Environmental Policy and Management also are available. Those who are interested in teaching at colleges and universities a master or doctorate degree after some experience could be necessary in these areas.   Many states recommend a certification or credentialing process for Conservation Scientists and

Foresters and have these opportunities available. Check with the academic facility of your choice for more information.

Attaining a master's degree is not normally a requirement to work in this career although many successful Conservation Scientists have earned masters and Ph.D. degrees.

When you graduate how many jobs may be available as a Conservation Scientist?

Employment of Conservation Scientists and Foresters is expected to increase by about seven percent (7%) through 2020. There has been about 34,000 Scientists in these areas in the past few years. Most of these scientists work for government agencies and private consulting organizations.

While earning degrees, many students participate in work-study programs while attending federal or state land grant colleges. Most soil conservation students earn degrees in agricultural science or hydrology with a concentration in water management. Others study wildlife biology or forestry.

The median annual salary for Conservation Scientists and Foresters is around $61,000. Some in this career earn less than $37,000 and the highest percent can earn around $105,000 annually. The projected increase of Conservation Scientists and Forester positions and a relatively small number employed in these sciences assists in making these careers **Highly Competitive**.

Interest in our ecosystems along with balancing human activities and preserving natural environments is leading the way for a slight increase in demand for Conservation Scientists and Foresters.

How can your Cognitive thinking Style and Skill Set assist in determining your success in working environments as a Conservation Scientist?

Successful Conservation Scientists have many practical abilities and Skill Sets **(Table 2)** along with excellent critical-thinking, analytical and observation skills and advanced mathematical ability. They are able to creatively interpret and apply data and information into projects, documentation and reports.

Practical and **Logical** Cognitive thinking **(Table 1)** styles can be very beneficial as a Conservation Scientist. Interest in nature preservation and balancing human needs with healthy and strong ecosystems are important skills for successful Conservation Scientists.

These scientists also possess the ability to creatively solve complex problems especially when determining the most effective route to balance natural diversity with human needs for land. Strong written skills, excellent communication skills and interpersonal skills are valued. They explain their findings and give recommendations.

They also give presentations to those in government, private and academic organizations. Successful Conservation Scientists possess excellent physical stamina while working in extreme weather conditions.

Working Environments **(Table 3)** in this career can range from **Highly Political (HP)** to **Least Political (LP).** In many academic and government positions working environments can be both **Highly Political** and **Bureaucracy/Conformity (BUR/CF)** can be valued and rewarded. Manufacturing facilities can also be **Highly Political** and **Bureaucracy (BUR/CF)** is usually valued. Research and Development areas within these organizations could be more open to **New Ideas (NI).**

Private research and development organizations could also be very **Highly Political** and in some organizations new methods are rewarded. Small or newly formed private working environments can be more open to new ideas **(NI).**

Conservation Scientists

Areas of Concentration

Many with careers in Conservation science have degrees and courses in similar areas of concentration including:

BIOCHEMISTRY   —   CELL   BIOLOGY — MICROBIOLOGY - Researching and analyzing cells, genes and molecular life.

ORGANIC CHEMISTRY - PHYSICAL CHEMISTRY - Organic chemistry involves researching and analyzing the properties and structure of plants and organisms and reactions.

QUANTUM BIOLOGY — Research in this area suggests plants at some level can recognize and maneuver to receive the most direct energy to assist in the photosynthesis process for growth and survival. Those in this area of concentration also research other organisms including biological.

MOLECULAR   BIOLOGY-   GENOMICS — MOLECULAR GENETICS - Molecular Biophysicists and Genomics research genetics and interactions between DNA, RNA and protein in living cells.

HYDROLOGY — Hydrologists research, examine and analyze data and information regarding the physical characteristics, distribution, and circulation of earths' water supply both on the surface and within the earth.

Careers in Conservation Science

There are three major areas of concentrations Conservation Scientists are employed within:

RANGE CONSERVATIONISTS

They protect and manage forests, grasslands, wetlands and even desert areas in the United States. Other titles in the concentration include **Range Managers**, **Range Scientists**, or **Range Ecologists**.

These scientists maintain and manage a full array of natural resources in the environment they are responsible for including vegetation, wildlife and water. Range Managers also research, discover and implement recovery solutions from natural disasters such as floods, invasive animal species, forest wildfires and other disasters. As a Range Manager responsibilities can include commercial and residential land development, recreational land areas, farming, preserving wildlife habitats, and mining in some states.

SOIL CONSERVATIONISTS

Soil conservationists assist landowners, farmers, ranchers within both private and government organizations. These scientists assist in developing and implementing effective strategies for land preservation while assisting to balance the needs of property owners along with the least impact to soil health. One of the major focuses of these scientists is to limit soil erosion in the use of land and property.

WATER CONSERVATIONISTS

Water Conservationists work to research, test and ensuring water supplies are free from pollution and contaminants.

They work with government organizations and also with land and property owners to preserve water resources.

There are a number of colleges and universities accredited by The Society of Range Management offering degrees in Range Management. **(See Information and Resources)**

ENVIRONMENTAL SCIENTISTS

What do they do?

Environmental Scientists use their knowledge of natural sciences to research, analyze and recommend environmental protection solutions. Collecting and analyzing environmental data they research safe water, food, soil and air plans and solutions. Environmental Scientists create investigations while discovering solutions to increase the health of the environment and eliminate pollution. Some of the plans and projects they develop have lead to eliminating pollution and restoring environments to healthy status.

Environmental Scientists provide information and plans to government and private organizations on environmental health risks and hazards. They also work along with construction organizations recommending low risk or the most effective solutions to limit or minimize environmental impact before and during construction projects. Environmental Scientists also work on restoring balance after natural disasters.

There are specific areas of concentrations for many Environment Scientists. One is as **Environmental Chemists** analyzing and researching the impact of chemicals within environments and ecosystems. Another is as Environmental **Health Specialists** analyzing and educating the public of potential health risks. **Environmental Protection Specialists** analyze and investigate human activity and impact on pollution levels along with other responsibilities.

Environmental Scientists recommend solutions regarding imbalances created by human activities including the depletion of the ozone layer issues. They also recommend and follow governmental guidelines. Other guidelines revolve around clean air, safe water and contaminant-free soil for plant growth and healthy ecosystems.

Environmental Scientists work with a 'big picture" viewpoint of a variety of factors affecting natural environments.   These factors include an ecological balance between human activities and natural resources including plants, trees, land, animals and soil preservation within these environments.

Where Do Environmental Scientists Work?

Some Environmental Scientists work for government organizations monitoring and ensuring compliance of guidelines and regulations. Many work for companies and corporations including consulting firms.

Other Environmental Scientists work in offices and laboratories within both government and corporations. They divide their time between working in the field and then back to offices or laboratories observing and analyzing collected data. Working in the field can be very physically demanding at times working in rain, snow and challenging weather conditions.

Some work for large consulting firms and engineering organizations with thousands of employees.   Others work within small firms working on long-term projects as they cooperate with other scientists and engineers.

These scientists work in a variety of specialties including Soil and Water Protection, and can involve a good amount of fieldwork in areas such as prairies and woodlands; many work in remote and hard to reach areas.

What type of degree do you need to become an Environmental Scientist?

Environmental Scientists should have a bachelor's degree in environmental science or in the natural sciences including biology, chemistry, or the earth sciences such as geology, hydrology, paleontology (the study of plant and animal fossils, past climates, decay and preservation) and petrology (the study of the origin, structure and process of rock formations).

Courses in Environmental Policy and Regulation and Waste Management have assisted many in this career. Those most successful in this field have earned either a master's degree or a Ph.D. A Ph.D. is necessary to teach at colleges and universities after years of experience in this field.

Strong written and speaking skills are important for most Environmental Scientists as they give presentations of their recommendations to those in government, private and academic organizations. Environmental Scientists can begin as field analyst or research assistants working in laboratories and offices. As experience is gained they can reach management or senior research positions.

It is important to choose a college or university that is not only an accredited school, but also one where your degree of choice is also accredited from that academic facility. Bachelor degrees in Environmental Science along with bachelor degrees in Natural Resources Ecology and Management are available. Colleges and universities in many states offer these degrees.

Courses for degrees in this area can include ecology and biology other science based courses. Masters in Environmental Policy and Management are available. Check with the academic facility of your choice.

When you graduate how many jobs may be available as an Environmental Scientists?

Employment of Environmental Scientists and Foresters is expected to increase by fifteen percent (15%) through 2020. This is more than double the percentage of expected employment increases for Conservation Scientists and Foresters. There has been about 88,000 Environmental Scientists in the past few years. Most of these scientists work for government agencies and private consulting organizations.

While earning degrees many students participate in work-study programs while attending federal or state land grant colleges. Most soil conservation students earn degrees in agricultural science or hydrology with a concentration in water management. Other studies include wildlife biology or forestry.

The median annual salary for Environmental Scientists is around $61,000. Some in this career earn less than $37,000 and the highest percent earn around $105,000 annually. The projected increase of Environmental Scientists positions and a relatively larger number of Environmental Scientists holding these positions assists in making these careers **Moderately Competitive**.

Interest in our ecosystems along with balancing human activities and preserving natural environments is leading the way for a higher than average increase in demand for Environmental Scientists.

How can your Cognitive thinking Style and Skill Set assist in determining your success in working environments as an Environmental Scientist?

Successful Environmental Scientist have many practical abilities and Skill Sets **(Table 2)** along with excellent critical-thinking, analytical and observation skills and advanced mathematical ability and are able to creatively interpret and apply data and information into projects, documentation and reports.

Practical and **Logical** Cognitive thinking **(Table 1)** styles can be very beneficial as an Environmental Scientist. A very high interest in nature preservation and balancing human needs with healthy and strong ecosystems are among important skills for successful Environmental Scientists.

These scientists also possess the ability to creatively solve complex problems especially when determining the most effective route to balance natural diversity with human needs for land. Very good communication and interpersonal skills are valued as they explain their findings. Successful Environment Scientists possess excellent physical stamina while working in extreme weather conditions.

Working Environments **(Table 3)** in this career can range from **Highly Political (HP)** to **Least Political (LP).** In government laboratories or regulating positions working environments can be both **Highly Political** and **Bureaucracy/Conformity (BUR/CF)** can be valued and rewarded. Manufacturing facilities can also be **Highly Political** and **Bureaucracy (BUR/CF)** is usually valued.   Research   and

Development areas within these organizations could be more open to **New Ideas (NI).**

Private research and development organizations could also be very **Highly Political** and in some organizations new methods are rewarded. Small or newly formed private working environments can be more open to new ideas **(NI).**

Environmental Scientists

Areas of Concentration

Many with careers in Environmental Science study courses and earn degrees in similar areas of concentration including:

BIOCHEMISTRY — CELL BIOLOGY — MICROBIOLOGY - Researching and analyzing cells, genes and molecular life.

ORGANIC CHEMISTRY - PHYSICAL CHEMISTRY - Organic chemistry involves researching and analyzing the properties and structure of plants and organisms and reactions.

HYDROLOGY — Hydrologists research, examine and analyze data and information regarding the physical characteristics, distribution, and circulation of earths' water supply both on the surface and within the earth.

There are some major areas of concentrations Environmental Scientists are employed within including:

Environment Scientists Careers

Environmental Chemists work specifically with ecosystems and the impact of chemicals on these systems. Some of the areas chemists work within are waste management and pollution and contamination control. Others research and analyze how acid can affect people, animals and plants.

Environmental Health Specialists research, analyze and educate the public of potential health risks and how environment pollutants impact human health. These risks include unsafe water and contaminated food products.

Environmental Protection Specialists analyze and investigate human activity and pollution levels in relation to preserving natural and man-made environments.

FOOD SCIENTISTS

What do they do?

Food Scientists work in many areas including developing and researching new and existing food products, understanding and improving food processes; researching genetics of plants; studying the biological and chemical makeup of food along with food packaging and design.

Food Scientists research, test, and monitor safe food and food products for humans; feed quality for animals; water quality and soil pollutant levels for plants and crops. Food, Agricultural and Animal Scientists have similar duties but also have distinct duties important to their specific discipline or area.

Food Scientists' duties are developed along a variety of areas including **Quality Assurance, Research and Development, Food Processing** and **Food Regulation**.

- **Food Scientists** working in the area of Quality Assurance includes monitoring food processes and production, analyzing ingredients and labeling of food products.
- A **Food Researcher** duty can involve experimenting with food storage methods and additives within food.
- **Food Developers** work includes researching new food products and improvements to existing food products.
- **Processing Food Scientists** duties can involve canning, drying, baking and pasteurization processes of food and food products.

- **Food Scientist Regulators** for the government can monitor and enforce food regulations.

Where Do Food Scientists Work?

Many **Food Scientists** work in government organizations and research and development within private and academic laboratories. Some Food and Agricultural Scientists work in the field providing hands-on research and others work for corporate organizations. A good amount of time is spent analyzing, studying and documenting data and reports. For many a high tolerance of a variety of environments is necessary due to regular visits to farms, manufacturing and food processing plants.

Some with bachelor's degrees in food and agricultural sciences find work in related jobs rather than becoming an agricultural or food scientist. A bachelor's degree in agricultural science is useful for managerial jobs in farm-related or agricultural inspection and ranch-related businesses. Organizations can include farm credit institutions along with businesses making or selling feed, fertilizer, seed, and farm equipment.

What type of education does it take to become a Food Scientist?

Some positions such as product developers require a bachelor's degree in agricultural science or related field. Advanced research requires attaining a Ph.D. With an advanced degree such as a doctorate (Ph.D.) in agricultural science courses could also involve plant genetic research and application. In order to teach at universities and college a masters degree or Ph.D. is needed.

In the U.S. most states have a land-grant college offering agriculture and food sciences curriculum and degrees.   Look for colleges offering degrees in agricultural science. A bachelor's degree in agricultural science can lead to work in farming, agricultural inspection or in companies manufacturing farming equipment. Undergraduate courses in biology, physics, engineering and chemistry can also lead to careers as an Agricultural Scientist.

When you graduate how many jobs may be available as a Food Scientist?

Employment of **Food** and **Agricultural Scientists** is expected to increase by about ten percent (10%) through 2020. High interest in food safety is fueling an increase in the need for more Food Scientists. This field does not employ large numbers of people. About 3,000 people are currently Food and Agricultural Scientists.

Most growth over the next ten (10) years for Food, Animal and Agricultural Scientists will be in private industry. Private industry has increased its demand for agricultural and food scientists for many reasons including a need for greater expertise in developing food, crops, and pharmaceutical items.

The projected increase of Food and Agricultural Scientists and a high demand for safe food and pharmaceutical products are good signs for future growth in this area. With all the above factored into the overall outlook makes this a **Highly Competitive** field.

The average salary after some work experience in the field can be about $46,000 for those working in state, local and private colleges and universities. With a Ph.D. those working in private organizations range from $57,000 or more annually.

A high demand to understand ecosystems and how pest infiltrate and impact both water and soil could lead to higher demands for these scientists. The need for renewal energy such as Biofuels derived from plant sources including plant seeds and sugar and other plant sources can lead to an increase in demand for Food Scientists.

How can your Cognitive thinking Style and Skill Set assist in determining your success in working environments as a Food Scientist?

Successful Food and Agricultural Scientists have abilities and Skill Sets **(Table 2)** such as excellent critical-thinking, analytical and observation skills and advanced mathematical ability and are able to creatively interpret and apply data and information into projects, documentation and reports.

If interested in designing and packaging **Creative** abilities are also highly valued in these areas. Food Scientists in this field who are successful have also mastered the "soft skills" of being a productive team player.   Having **Practical** and **Logical** Cognitive thinking styles **(Table 1)** can be very beneficial as a Food Scientist.

Working Environments **(Table 3)** in this career can range from **Highly Political (HP)** to **Least Political (LP).** In many academic and government positions working environments can be both **Highly Political** and **Bureaucracy/Conformity (BUR/CF)** can be valued and rewarded. Manufacturing facilities can also be **Highly Political** and **Bureaucracy (BUR/CF)** is usually valued.   Research and Development areas within these organizations could be more open to **New Ideas (NI).**

Private research and development organizations could also be very **Highly Political** and in some organizations new methods are rewarded. Small or newly formed private working environments can be more open to new ideas **(NI).**

Food Scientists

Areas of Concentration

FOOD CHEMISTRY – Those in this area research and analyze the molecular components and structure of food along with understanding how various foods react to chemicals.

FOOD ENGINEERING – These scientists are involved in research and development along with industrial processes used to manufacture food products.

FOOD MICROBIOLOGY, FOOD TECHNOLOGY, MOLECULAR GASTRONOMY, FOOD SAFETY AND PRESERVATION – These scientists research, test and prevent microorganisms influencing the safety of foods including preventing food poisoning, spoilage and food-borne diseases like salmonella. They work to discover safe food preservation. These scientists can also work on creating food safety guidelines with government organizations responsible for food safety.

FOOD PACKAGING AND DESIGN – These scientists design and create the most effective packaging for a variety of foods.

FOOD PHYSICIST AND SENSORY ANALYSIS – These scientists use experimental design and statistical analysis to understand the types of qualities the consumer prefer in food. They can research food in relationship to consumers' preference regarding texture, appearance, taste and smell.

FOOD PRODUCT DEVELOPMENT – These scientists research and develop how to improve the flavor and nutritional value of food.   They can also work on improving manufacturing techniques to improve the look and texture of food products.   Their research assists in introducing new food products to consumers.

Food Scientist Careers

Food Scientist Careers include a variety of titles. Organizations vary regarding necessary degree requirements and work environments. The information below is provided for only general information.   Look into your area of interest for degree information.

SPECIALTY   DEGREE NEEDED

Agricultural Inspection and Research
Agricultural Scientists   Bachelors and Higher
Food Scientists   Bachelors and Higher
Government Regulators   Masters and Higher
Product Developers   Bachelors and Higher
Soil Scientists   Bachelors and Higher

Academic Institutions
Academic Instructors   Masters or (PH.D.)
Academic Researchers   Doctorate (PH.D.)

Manufacturing
Agricultural Inspector   Masters and Higher

FORESTERS

What do they do?

Foresters' concentration is in the area of forest and tree conservation and survival. **Foresters** and Conservation Scientists perform very similar work. Both **Foresters** and Conservation Scientists work to research, understand, manage and implement ecologically safe relationships between nature and humans. **Foresters** analyze data on soil quality, insect and diseases damaging trees and plants.

Foresters monitor and enforce environmental protection regulations. **Foresters** monitor adverse conditions resulting in fire hazards to trees and plants. Foresters also coordinate and implement fire control protection and activities among forestry workers. Foresters are responsible for being aware of the quality and quantity of trees and for maintaining a diversity of a variety of tree species.

Foresters analyze, track and monitor corporations, private owners and loggers in their use of timber for activities and use. They assist in building roads, pathways and trails within forested areas. They oversee campsites and recreational facilities with an eye toward safety for plants, trees, wildlife and humans.

Foresters can be responsible for coordinating new seedlings planting to refurbish and maintain healthy forest areas. They implement measures for tree health by pinpointing and assisting in removing diseased trees. They are also responsible for public awareness of ongoing forestry and conservation programs and activities.

Foresters are similar to Conservation Scientists who inform and assist private landowners of forestry property lines while keeping in mind human activities regarding land ownership and areas used for recreation.

Foresters work with a 'big picture" viewpoint of a variety of factors affecting natural environments. These factors include an ecological balance between human expansion and activities and natural resources including plants, trees, land, animals and forest and soil preservation within these environments.

Foresters work closely with landowners, farmers and property owners allowing natural land preservation meeting the needs of property owners.

These scientists make use of specific knowledge while developing conservation strategies to maintain biological diversity.   They also work with residential and commercial property owners for recreational purposes.

Where Do Foresters Work?

Most Foresters work for government organizations. Some **Foresters** work for private organizations. Other Foresters and Conservation Scientists work with advocacy organizations interested in maintaining ecological balance and diversity within nature and human activities.   With experience some work independently.

Careers in government include performing studies regarding areas impacted by human expansion and activities; other studies can include researching and recommending ecological benefits of re-forestation in negatively impacted areas.   Private organizations including private landowners may use Foresters regarding private and public property lines and forest areas.

These scientists work in a variety of specialties including Soil and Water and Forest Protection, and can involve a good amount of fieldwork in areas such as prairies and woodlands; many work in remote and hard to reach areas. Some Foresters also work in offices.

What type of degree do you need to become a Forester?

Research available options to ensure the school you are attending or plan to attend is an accredited school in the field of your study and degree attainment. Most Foresters have earned a bachelor's degree along with higher degrees in forestry, agriculture, environmental science or rangeland management.

Bachelor degrees in Environment Science along with bachelor degrees in Natural Resources Ecology and Management are available. Colleges and universities in many states offer forestry degrees. Courses for degrees in this area can include Forest Resource Management, ecology, biology and other courses.

Those most successful in this field have earned either a master's degree or a Ph.D. Attaining a master's degree is not normally a requirement to work in this career although many Foresters have earned masters and Ph.D. degrees. A Ph.D. is necessary to teach at colleges and universities after years of experience in this field.

Strong written and speaking skills are important for Foresters as they give presentations of their recommendations to those in government, private and academic organizations.

Masters in Environmental Policy and Management also are available. Many states recommend a certification or credentialing process for Foresters and have these opportunities available.   Check with the academic facility of your choice for more information.

When you graduate how many jobs may be available as a Forester?

Employment of Foresters and Conservation Scientists is expected to increase by seven percent (7%) through 2020. About 34,000 people are employed as Foresters and Conservation Scientists. Decreases in local, state and federal funds and budgets have greatly impacted funds allocated to conservation. Many Foresters work within the rangelands in the United States in the western states and in the State of Alaska.

A relatively small number of people out of the total working population are Foresters and Conservation Scientists. With a relatively small base of current and potential workers in this field along with smaller amounts of available funding makes this a **Highly Competitive** field.

Each state has colleges or universities offering accredited degrees in Range Management. The Society of Range Management accredits many colleges and universities.   The Society of American Foresters also accredits many programs offered in forestry and conservation. **(See Information and Resources)**

While earning degrees, many students participate in work-study programs offered by federal or state land grant colleges. If Range Management is your interest degrees in range management and range science are offered at some colleges and universities. Range management courses include animal, plant and soil sciences along with ecology and resource management studies.

Some soil conservation students earn degrees in agricultural science or hydrology with a concentration of water management. Others study wildlife biology or forestry. The median annual salary for Foresters is around $55,000. Some in this career earn less and the highest percent earn around $75,000 annually.

Interest in our ecosystems along with balancing human activities and preserving natural environments is leading the way for an increase in the demand for Foresters, Conservation and Environmental Scientists.

How can your Cognitive thinking Style and Skill Set assist in determining your success in working environments as a Forester?

Successful Foresters have many practical abilities and Skill Sets **(Table 2)** along with excellent critical-thinking, analytical and observation skills and advanced mathematical ability and are able to creatively interpret and apply data and information into projects, documentation and reports.

Practical and **Logical** Cognitive thinking **(Table 1)** styles can be very beneficial as a Forester. A very high interest in nature preservation along with keen observation skills to detect small changes and patterns in natural environments are some of the most important skills for successful Foresters.

These scientists also possess the ability to creatively solve complex problems especially when determining the most effective route to balance natural diversity with human needs for land.   Very good communication and interpersonal skills are valuable as these scientists explain their findings to peers, management and to the public.   Successful Foresters possess excellent physical stamina while working in extreme weather conditions.

Working Environments **(Table 3)** in this career can range from **Highly Political (HP)** to **Least Political (LP).**   In many academic and government positions working environments can be both **Highly Political** and **Bureaucracy/Conformity (BUR/CF)** can be valued and rewarded.   Manufacturing facilities can also be **Highly Political** and **Bureaucracy (BUR/CF)** is usually valued.   Research and Development areas within these organizations could be more open to **New Ideas (NI).**

Private research and development organizations could also be very **Highly Political** and in some organizations new methods are rewarded.   Small or newly formed private working environments can be more open to new ideas **(NI).**

Foresters

Areas of Concentration

Many with careers in Forestry also have skills and degrees and courses in similar areas of concentration including:

BIOCHEMISTRY — CELL BIOLOGY — MICROBIOLOGY - Researching and analyzing cells, genes and molecular life.

ORGANIC CHEMISTRY - PHYSICAL CHEMISTRY - Organic chemistry involves researching and analyzing the properties and structure of plants and organisms and reactions.

HYDROLOGY — Hydrologists research, examine and analyze data and information regarding the physical characteristics, distribution, and circulation of earths' water supply both on the surface and within the earth.

Careers as Foresters

Other titles in this area include **Range Managers**, **Range Scientists**, or **Range Ecologists**.

RANGE CONSERVATIONISTS

They protect and manage forests, grasslands, wetlands and even desert areas in the United States.

These scientists maintain and manage a full array of natural resources in the environment they are responsible for including vegetation, wildlife and water. Range Managers also research, discover and implement recovery solutions from natural disasters such as floods, invasive animal species, forest and forest wildfires. As a Range Manager areas can span over many areas including commercial and residential land development, recreational land areas, farming along with mining in some states, or wildlife preservation and habitats.

SOIL CONSERVATIONISTS

Soil conservationists assist landowners, farmers, and ranchers including government agencies.   These scientists assist in developing and implementing effective strategies for land preservation while balancing the needs of property owners with the least impact to soil health.   One of the major focuses for these scientists is to limit soil erosion in the use of land and property.

WATER CONSERVATIONISTS

Water Conservationists work to research, test and ensure that water supplies are free from pollution and contaminants. They work with government agencies and also with land and property owners to preserve water resources.

VETERINARIANS

What do they do?

Veterinarians diagnose and care for the health of animals. They research medical conditions of animals including pets, zoo animals, horses and livestock. Veterinarians also vaccinate animals against diseases, perform surgery on animals when necessary and prescribe medication among a variety of duties.

There are a number of careers for those interested in Veterinary care. Those who assist with the medical needs and health of **Pets** are also called **Companion Animal Veterinarians** and usually work in private clinics assisting animals with medical and health issues and concerns as medical doctors do for humans.

A small number of **Veterinarians** also work exclusively with horses and are called **Equine Veterinarians.** Others in this field are **Food Safety and Inspection Veterinarians** caring for the health of farm animals and livestock. **Research Veterinarians** work in laboratories performing tests and researching the best medicines for animals. They analyze how to avoid illnesses and diseases for these animals.

Where Do Veterinarians Work?

Most Veterinarians are **Companion Animal Veterinarians** and work in private clinics or veterinarian medical offices. At times some **Companion Animal Veterinarians** who care for pets may have to euthanize animals. Veterinarians who care for pets are also called **Clinical Veterinarians**.

Equine Veterinarians treat horses and livestock traveling between their offices and to farms or facilities and often work outdoors in a variety of weather conditions.   They may perform emergency surgery when necessary on farm grounds, outdoors or in stables.   These Veterinarians travel between their offices and farm facilities.

Food and Safety Veterinarians travel to food processing plants, slaughterhouses and farms inspecting food safety.

Food Animal Veterinarians treat farm animals such as cattle, pigs and sheep.

Research Veterinarians spend a good amount of time working in laboratories. They also advise those in private and government organizations. They may have little contact with animals.

Most Veterinarians work with sick, injured and diseased animals and this work can be emotionally stressful at times.   Sick animals can be noisy and unpredictable; care must be taken to avoid being bitten or other injuries to the Veterinarians. Most Veterinarians usually work long hours including nights and weekends as they respond to emergencies.

What type of education does it take to become a Veterinarian?

To become a Veterinarian earning a Doctor of Veterinary Medicine (D.V.M.) degree or a Veterinary Medicine degree (V.M.D.PhD) is necessary. The Doctor of Veterinary Medicine degree (D.V.M.) is more widespread in the United States. The Veterinary Medicine degree (V.M.D.PhD) is not offered at all veterinary colleges. These degrees are similar and the required coursework must be completed from an accredited veterinary college. Currently there are more than twenty accredited veterinary colleges in the U.S. offering these programs. A program through one of these colleges offering classes, laboratory work and clinical studies takes four years (4) to complete. About 2,500 students graduate each year from veterinarian colleges.

Although a bachelor's degree is required before attending a veterinary college it is not necessary to have earned a four-year bachelors degree at one of the accredited veterinary colleges. If earning a bachelors degree at a college or university other than a veterinary college and to increase the likelihood of matching the curriculum required at veterinary schools take required courses including animal science, anatomy, biology, chemistry, microbiology, physiology, wildlife biology and zoology. Other important courses include mathematics, and humanities or social sciences to understand behavior. Contact the schools of your choice for more specific information.

After a bachelor's degree is earned programs at veterinary colleges are four years with three years of clinical, laboratory and classroom work. The fourth year of the program most students work on clinical rotations at a veterinary medical facility.

Relatively new to the courses taken by those attending veterinary colleges are business courses assisting with accounting and business finance for private practices. Look into the veterinary program of interest for requirements.

Admission to veterinary colleges is very competitive where less than half that apply are admitted. To have the best advantage for acceptance into these programs at veterinary colleges having prior experience working with animals, even as a part-time job can be beneficial.

Those having some experience working with animals as a student can find this experience useful when applying at most veterinary medical colleges. Other experience can include working with Veterinarians in clinics. Those with hands-on experience working with animals as interns in the farming and manufacturing areas could also have a slightly greater advantage. More experience can include working on farms or horse stables or animal shelters.

To work as a veterinarian in the United States a license is required. Licensing requirements vary state by state. Those seeking to become a Veterinarian must first complete an accredited veterinary program and then pass a Veterinary Licensing Exam.

A national veterinary exam with a passing score is required for all practicing Veterinarians. Each state requires a state-level exam be passed for that particular state. Most states require both a state and national exam be passed. State level licenses are not transferable from one state to another.

After receiving a degree from an accredited veterinary college and passing the appropriate state and national veterinary exams a graduate can practice as a Veterinarian. Some new Veterinarians can enter a one-year internship program to gain some experience.

There are also certification programs available. After earning a Doctor of Veterinary Medicine (D.V.M.) and passing the state and national licensing exams and after actually working as a Veterinarian, certifications are available although they are not required. Certifications can assist in remaining current with new information and techniques in the field.

When you graduate how many jobs may be available as a Veterinarian?

Employment of Veterinarians in all disciplines is expected to increase by thirty six percent (36%) through 2020. This area is expected to grow faster than most careers.

There has been about 61,000 Veterinarians in the past few years.   Most Veterinarians work in veterinary services clinics and organizations.   Over seventy percent (70%) work in private clinical practices treating pets.   Others work within government organizations, colleges and universities or at private medical or research laboratories.

Overall this is a **Moderately Competitive** field due in part to the demand for Veterinarians is increasing and this field is expected to increase more than thirty percent (36%).   The **Companion Animal Veterinarians** field (those working with Pets) is the most **Highly Competitive** in part because many are interested in careers in this area while a limited number are accepted into Veterinary programs.

Recently the annual average salary of Veterinarians is about $80,000. The lowest salary of Veterinarians was as $46,000 and the highest salaries were about $144,000 and above.   Those working in federal Veterinarian positions could earn more than those working in other organizations.

Some who have finished their degrees and veterinarian requirements are being highly sought after by both government and private organizations. Many are working in high demand areas of food and safety regulations and pharmaceuticals researching new drugs and medicines.

Most in Veterinary colleges prefer to concentrate in the area of **Companion Animal Veterinarians** (also called **Clinical Veterinarians**). A growing pet population is driving the need for more of these Veterinarians.

Currently pet owners consider their pets a member of their family and are willing to pay the costs for healthy pet care. Some of these costs include relatively expensive medical treatments to improve their pets' health. Veterinarian medicine and procedures are also highly advanced similar to treatments offered to humans such as cancer treatments and organ transplants.

A growing human population is also fueling the need for more **Food and Safety Veterinarians** due to greater concern for animal safety, healthy and safe products and disease control. More of these Veterinarians will be needed to ensure animal and human food supply is safe.

How can your Cognitive thinking Style and Skill Set assist in determining your success in working environments as a Veterinarian?

Successful Veterinarians have a variety of Skill Sets **(Table 2)** along with high interests in animal health and safety. Compassion and understanding are also very important abilities for Veterinarians especially those who care for pets. Veterinarians should master dexterity skills as they treat small and large animals.

Practical and **Logical** Cognitive thinking **(Table 1)** styles can be very beneficial as a Veterinarian. All Veterinarians should also have excellent <u>critical-thinking</u> and <u>mathematical ability</u>. Excellent <u>observation </u>and <u>problem solving </u>skills are necessary to diagnose illnesses and decide on treatments. Successful Veterinarians have also mastered the "soft skills" of communicating with owners and peers. Management skills are important for those operating and managing their own private laboratories or clinics.

Working Environments **(Table 3)** in this career can range from **Highly Political (HP)** to **Least Political (LP).** In many academic and government positions working environments can be both **Highly Political** and **Bureaucracy/Conformity (BUR/CF)** can be valued and rewarded. Manufacturing facilities can also be **Highly Political** and **Bureaucracy (BUR/CF)** is usually valued.   Research and Development areas within these organizations could be more open to **New Ideas (NI).**

Private research and development organizations could also be very **Highly Political** and in some organizations new methods are rewarded. Small or newly formed private working environments can be more open to new ideas **(NI).**

Veterinarians

Areas of Concentration

COMPANION ANIMAL VETERINARIANS usually work in private clinics treating and caring for pets including cats, dogs, birds, ferrets, rabbits and other pets. These veterinarians are also called **CLINICAL VETERINARIANS.** Along with diagnosing and treating animal medical problems these Veterinarians also consult and advise the animals' owners. Most perform medical procedures, set broken bones and fractures and give vaccinations among a variety of duties.

EQUINE VETERINARIANS make up a small percentage of Veterinarians and usually spend their time treating and maintaining the health of horses.

FOOD ANIMAL VETERINARIANS treat farm animals such as cattle, pigs and sheep. A small percentage of Veterinarians work exclusively as Food Animal Veterinarians. Their time is spent between farms and ranches. They test and vaccinate animals and advise and counsel owners regarding general animal health practices.

FOOD SAFETY AND INSPECTION VETERINARIANS usually work for government organizations inspecting livestock, feed and animal products while enforcing safety regulations. Some of these Veterinarians research preventable methods to avoid and eliminate a variety of diseases and viruses. They also inspect facilities for proper sanitation procedures and processing including manufacturing plants processing animal food products.

RESEARCH VETERINARIANS apply their skills and focus on testing animals and conducting research while discovering and identifying solutions to both animal and human health issues. Some research pharmaceuticals and drugs assisting in controlling or eliminating illnesses and diseases.

Some Veterinarians become **INSTRUCTORS** and teach at colleges and universities.

WILDLIFE BIOLOGISTS

What do they do?

Wildlife Biologists research, analyze and study animals in natural settings and habitats and in controlled settings to understand animal behavior, disease and pollution controls and ecological balance. Wildlife Biologists research, analyze and study living organisms. They study how these organisms including animals interact with and influence their environments including controlled environments such as zoos and designated preservation areas and parks.

Wildlife Biologists study animal's natural characteristics in the wild. They research and analyze species and their interactions with same and different species in the wild and within controlled environments. They research the nutrition, movement, activities and reproduction within various species. These scientists research and analyze animal diseases, reproduction and natural activities and patterns.

Some results of the research these scientists perform include estimating wildlife population size and the variety of species. Wildlife Biologists and Zoologists are instrumental in researching the various species threatened in the wild.

These scientists collect biological and behavioral data of daily activities of animals for analysis and writing research papers and reports. These scientists also analyze and give findings regarding how humans influence natural wildlife and habitats and provide information regarding wildlife conservation and management.

Geographical information analyzed by these scientists track animal behavior for a variety of reasons including threats to wildlife and the spread of invasive species. They also monitor animal populations working to ensure these populations remain at sustainable levels. Some Zoologists and Wildlife Biologists work with Environmental Scientists and Hydrologists to track and monitor water pollution levels and its impact on fish populations.

Where do Wildlife Biologists Work?

Wildlife Biologists work in a variety of areas. Some are **Researchers** who use data to solve animal health problems among various duties. They also study, collect and catalog specimens. Other Wildlife Biologists are **Wildlife Educators** analyzing, researching and exploring wildlife behavior and reporting their findings in papers and giving presentations. **Wildlife Rehabilitators** provide care for injured, sick or orphaned wildlife with a goal of releasing them back into their habitats.

There are two major categories of **Wildlife Biology Researchers.** One includes **basic research** assisting in human understanding of wildlife issues and solutions. The other is **applied research** which is geared toward solving specific problems including disease control in animal populations.

Wildlife Biologists involved with basic research write grant proposals for funding from private organizations, colleges, universities and federal government organizations.

The National Science Foundation and the National Institutes of Health are among organizations awarding grants for wildlife biology research. These organizations look for new discoveries and ideas that can be put to practical use. **(See Information and Resources)**

Those in the applied research area supply information resulting in knowledge leading to avoidance of some specific human and animal illnesses and diseases. Other results of applied research include creating new pharmaceuticals, treatments and diagnostic measures. Some of this research has also assisted in developing new Biofuels.

Those who work in the applied research area usually work for private organizations and companies researching specific products for production and marketing. Having excellent writing, presentation and people skills is important. These scientists present their findings to managers and peers and also to decision makers.

Some Wildlife Biologists are also working along with biotechnologists who research and analyze the genetic levels of plants and animals to assist in understanding disease control. Fieldwork in this area can be physically demanding.

Most Wildlife Biologists work for colleges and universities as researchers and teachers. Government organizations employ a large number of Wildlife Biologists in the areas of wildlife management, conservation and also in agricultural areas. A few Wildlife Biologists work within private organizations.

What type of education does it take to become a Wildlife Biologist?

The minimum academic degree which someone interested in becoming a Wildlife Biologists should earn is a bachelor's degree for an entry-level position. For advancement in this field a master's or a doctorate (Ph.D.) is more often necessary. As an independent study researcher or to teach at a college or university a Ph.D. is required in this area.

A variety of colleges and universities offer bachelors degrees in wildlife biology or zoology. Others offer degrees in ecology a discipline closely related to wildlife biology. To earn a bachelors degree courses can involve anatomy, ecology, cellular biology and wildlife management. Courses can also concentrate on specific groups of animals such as amphibians or mammals.

They also take courses focusing on particular groups of animals, such as ichthyology (fish) or ornithology (birds). Courses in botany, chemistry, and physics are important because zoologists and wildlife biologists must have a well-rounded scientific background. Mathematical, statistical and Computer Science courses are also very important for Wildlife Biologists as they perform complex data analysis.

When you graduate how many jobs may be available as a Wildlife Biologist?

Employment of Wildlife Biologists and Zoologists in all the areas of concentration is expected to increase by about seven percent (7%) through 2020. About 20,000 Wildlife Biologists and Zoologists currently hold these positions. This growth is slower than the average for most careers and employment can be somewhat of a challenge to attain.

The majority of Wildlife Biologists work for federal and state colleges, universities and government organizations. Some also work for private organizations. Local, state and federal budgets are a factor regarding the number of Wildlife Biologists to be employed. Small levels of opportunities for new Wildlife Biologists and a relatively small number of people working in this field make this field **Highly Competitive**.

More interest is revolving around understanding the impact of humans on wildlife and their habitat. As human population grows and more threats to wildlife such as invasive species, diseases, habitat losses, human impact and climate changes creates a need for more Wildlife Biologists.

The median annual salary for Wildlife Biologists and Zoologists is around $58,000. Some in this career earn less than $36,000 and the highest percentage can earn around $94,000 annually.

Biofuels derived from plants and other renewal energy sources are also driving a need for more Wildlife Biologists. Interest in and demand for sustainable practices such as using animal waste instead of chemicals to naturally sustain environments has also increased.

How can your Cognitive thinking Style and Skill Set assist in determining your success in working environments as a Wildlife Biologist?

Successful Wildlife Biologists have abilities and Skill Sets **(Table 2)** including excellent <u>critical-thinking,</u> <u>analytical</u> and <u>observation</u> <u>skills</u> and <u>advanced</u> <u>mathematical</u> <u>ability</u> and are able to creatively interpret and apply data and information into projects, documentation and reports. Having a high interest in understanding natural ecosystems and wildlife is important in this field. Keen interests in ecosystems drive many in this field. As a Wildlife Biologist it is important to possess well-rounded knowledge within a scientific background.

They have also mastered the "soft skills" of being a productive team player. Having Practical and Logical Cognitive thinking styles **(Table 1**) are very beneficial as a Wildlife Biologist as they work with scientists in a variety of fields.

Working Environments **(Table 3)** in this career can range from **Highly Political (HP)** to **Least Political (LP).** In many academic and government positions working environments can be both **Highly Political** and **Bureaucracy/Conformity (BUR/CF)** can be valued and rewarded. Manufacturing facilities can also be **Highly Political** and **Bureaucracy (BUR/CF)** is usually valued. Research and Development areas within these organizations could be more open to **New Ideas (NI).**

Private research and development organizations could also be very **Highly Political** and in some organizations new methods are rewarded. Small or newly formed private working environments can be more open to new ideas **(NI).**

Wildlife Biologists

Areas of Concentration

Some **Wildlife Biologists** are also **Researchers** analyzing animal health problems, collecting species data in natural and controlled environments.   They perform experiments, collect specimens, write reports and record findings.   They can also give presentations and lectures along with preparing grant proposals.

Wildlife Educators analyze, research and explore wildlife behavior on location within natural habitats or in controlled environments.   They produce reports for papers, printed materials and giving presentations.

Wildlife Rehabilitators care for injured, sick or orphaned wildlife for eventual release back into their natural surroundings.

Wildlife Biologists Careers by Species

Many Wildlife Biologists and Zoologists specialize by concentrating their work on one species.   Some of those are below:

ENTOMOLOGISTS research and analyze insects. These scientists research different species and their numbers in the wild.

ECOLOGISTS research and analyze a variety of organisms within their environments including animals and insects to determine their relationship to healthy ecosystems.

EVOLUTIONARY BIOLOGISTS analyze and study the origins of a variety of species and observe how they have evolved over time.

HERPETOLOGISTS research and study snakes, frogs, amphibians and reptiles.

ICHTHYOLOGISTS identify fish and observe their behavior and provide their findings in reports and also publish and present scientific papers.

LIMNOLOGISTS research organisms living in fresh water. They research evolution and observe behavior.

MAMMALOGISTS study evolution and function of mammals in the wild and within controlled environments. Mammals are researched, analyzed and studied includes mammals giving live births such as monkeys, apes, whales, elephants and the rare mammals which lay eggs such as platypus.

MARINE BIOLOGISTS research functions, structure and the evolution of animals and organisms living in saltwater.

ORNITHOLOGISTS scientifically study birds. Many in this field have degrees in other fields such as ecology or biology.

Scientists with Similar Disciplines

BIOCHEMISTS – Research and study large and small organisms including animals.

MICROBIOLOGISTS – Research and study bacteria, algae and fungi including how they grow and reproduce.

PHYSIOLOGISTS – Research and study entire life cycles of animals and plants.

ZOOLOGISTS

What do they do?

Zoologists research, analyze and study animals in natural settings and habitats and in controlled settings with objectives to understand animal behavior, disease and pollution controls and ecological balance. Only a few Zoologists are also Zookeepers whose primary responsibilities are to care for and monitor a variety of animals usually in a controlled environment such as a zoo.

Zoologist's careers encompass a wide variety of duties and responsibilities. Most Zoologists research, analyze and study living organisms and how these organisms including a variety of animals, interact with and influence their environments.

Zoologists study animal's natural characteristics in the wild and in controlled environments such as designated preserve areas and zoos. They research and analyze species and their interactions with different species in the wild and controlled environments.

These scientists research and analyze animal diseases, reproduction and natural activities and patterns; and how humans influence natural habitats. They estimate wildlife population size and the variety of species. Zoologists collect biological data of animal activities for analysis; write research papers and provide their findings.

Geographic information analyzed by these scientists track animal behavior for a variety of reasons including threats to wildlife and the spread of invasive species.

They also monitor animal populations working to ensure these populations remain at sustainable levels.   Some Zoologists   and   Wildlife   Biologists   work   with Environmental Scientists and Hydrologists to track and monitor water pollution levels and its impact on fish populations.

Where Zoologists Work

A few Zoologists work as **Zookeepers** caring for, grooming   and   training   animals   in   controlled environments such as zoos.   Most Zoologists use a portion of their time in laboratories analyzing and researching information regarding various animals. Zoologists   research   various   diseases   and   their underlying causes with goals of discovering solutions.

Some   work   on   scientific   studies   and   observation regarding animal behavior.   They may breed various animals to understand mating patterns and behavior within family units. Many museums and Zoos employ Zoologists.

Some Zoologists are **Researchers** using collected data to solve animal health problems. They also analyze and catalog information and data on various specimens. Zoological Researchers and some Wildlife Biologists study both live animals and specimens for scientific research and application.

Other Zoologists are **Wildlife Educators** analyzing, researching and exploring wildlife behavior, reporting and giving presentations on their findings.   **Wildlife Rehabilitators** provide care for injured, sick or orphaned wildlife with goals of releasing them back into their habitats.

There are two major categories of zoology and wildlife biology researchers. One is **basic research** assisting in human understanding of wildlife issues and solutions. The other is **applied research** geared toward researching and solving specific problems including disease control in animal populations. Applied research could result in providing knowledge to avoid some specific human illnesses and diseases. Research in these categories has lead to creating new pharmaceutical drugs, treatments and diagnostic measures.

Those in basic research write grant proposals for funding from private organizations, colleges, universities and federal government organizations. The National Science Foundation and the National Institutes of Health are among organizations awarding grants for zoology and wildlife biology research.   These organizations look for new discoveries and ideas that can be put to practical use. **(See Information and Resources)**

Those working in the applied research area can usually work for private organizations and companies researching specific products for production and marketing. Having excellent writing, presentation and people skills is important here as some of these scientists present their findings to decision makers.

Some Zoologists are also working along with biotechnologist who research and analyze the genetic levels of plants and animals to assist in disease maintenance and control. Fieldwork in this area can be physically demanding.

Most Zoologists work for colleges and universities as researchers and teachers. Government organizations employ a large number of Zoologists and Wildlife

Biologists in the areas of wildlife management, conservation and also in agricultural areas.   Some Zoologists work within private organizations.

What type of education does it take to become a Zoologist?

For an entry-level position, the minimum academic degree for someone interested in becoming a Zoologist should earn is a bachelor's degree. For advancement in this field master's or doctorate (Ph.D.) degrees are more often necessary.   As an independent study researcher or to teach at a college or university a Ph.D. is required in this area.

A variety of colleges and universities offer bachelors degrees in wildlife biology or zoology. Others offer degrees in ecology a discipline closely related to Zoology and Wildlife Biology.   To earn a bachelors degree courses can involve anatomy, ecology, cellular biology, botany, chemistry and wildlife management. Courses can also concentrate on specific groups of animals such as Herpetology (reptiles) or Mammalogy (mammals).

As a Zoologist and Wildlife Biologist it is important to possess well-rounded knowledge with a scientific background.   Mathematics, statistical work and Computer Science courses are also very important for Zoologists and Wildlife Biologists to perform complex data analysis.

When you graduate how many jobs may be available as a Zoologist?

Employment of Zoologists and Wildlife Biologists in all the areas of concentration is expected to increase by seven percent (7%) through 2020 for both Zoologists and Wildlife Biologists. About 20,000 Zoologists and Wildlife Biologists currently hold these positions. Those who are successful as Zoologists excel in the biological sciences and mathematics. The majority of Zoologists work for federal and state colleges, universities and government organizations.

Some also work for private organizations. Small levels of opportunities for new Zoologists and a relatively small number of people working in this area make this field **Highly Competitive**.

More interest is revolving around understanding the impact of humans on wildlife and their habitat. As human population grows and more threats to wildlife occurs including invasive species, diseases, habitat losses and climate changes more Zoologists will be needed. Unfortunately smaller local, state and federal budgets are a factor regarding the number of Zoologists who are to be employed.

The median annual salary for Zoologists is around $58,000. Some in this career earn less than $36,000 and the highest percent can earn around $94,000 annually.

A high demand to understand human impact on animal habitats could lead to greater demands for more Zoologists and Wildlife Biologists. Biofuels derived from plants and other renewal energy sources are also driving a need for more Zoologists. Interest in and demand for sustainable practices such as using animal waste instead of chemicals to naturally sustain environments has also increased.

How can your Cognitive thinking Style and Skill Set assist in determining your success in working environments as a Zoologist?

Successful Zoologists have abilities and Skill Sets **(Table 2)** such as excellent <u>critical-thinking</u>, <u>analytical</u> and <u>observation</u> <u>skills</u> and <u>advanced</u> <u>mathematical</u> <u>ability</u> and are able to creatively interpret and apply data and information into projects, documentation and reports. Having a high interest in understanding animal biology and behavior, natural ecosystems and wildlife is important in this field. A keen interest of understanding the balance of humans and wildlife drive many in this field.

They have also mastered the "soft skills" of being a productive team player. Having **Practical** and **Logical** Cognitive thinking styles **(Table 1)** can be very beneficial as a Zoologist.

Working Environments **(Table 3)** in this career can range from **Highly Political (HP)** to **Least Political (LP).** In many academic and government positions working environments can be both **Highly Political** and **Bureaucracy/Conformity (BUR/CF)** can be valued and rewarded. Manufacturing facilities can also be **Highly Political** and **Bureaucracy (BUR/CF)** is usually valued. Research and Development areas within these organizations could be more open to **New Ideas (NI).**

Private research and development organizations could also be very **Highly Political** and in some organizations new methods are rewarded. Small or newly formed private working environments can be more open to new ideas **(NI).**

Zoologists

Areas of Concentration

Zookeepers' most important duties are to ensure animals in their care are healthy. They exercise and train animals along with preparing and monitoring meals. They groom animals and clean enclosures. They observe and record behavior of animals in their care. They explain behavior and care of animals to groups.

Some **Zoologists** are also **Researchers** analyzing animal health problems, collecting species data in natural and controlled environments. They perform experiments, collect specimens, write reports and record findings. They can also give presentations and lectures along with preparing grant proposals.

A few **Zoologists** work as **Wildlife Educators** analyzing, researching and exploring wildlife behavior on location within the natural habitat or in controlled environments. They produce reports including papers and printed materials and giving presentations.

Other **Zoologists** work as **Wildlife Rehabilitators** caring for injured, sick or orphaned wildlife for eventual release back into their natural surroundings.

Zoologists and Wildlife Biologists Careers By Species

Many Zoologists and Wildlife Biologists specialize by concentrating their work on one species. Some of those are below:

ENTOMOLOGISTS research and analyze insects. These scientists research different species and their numbers in the wild.

ECOLOGISTS research and analyze a variety of organisms within their environments including animals and insects to determine their relationship to healthy ecosystems.

EVOLUTIONARY BIOLOGISTS analyze and study the origins of a variety of species and observe how they have evolved over time.

HERPETOLOGISTS research and study snakes, frogs, amphibians and reptiles.

ICHTHYOLOGISTS identify fish and observe their behavior and provide their findings in reports and also publish and present scientific papers.

LIMNOLOGISTS research organisms living in fresh water. They research evolution and observe behavior.

MAMMALOGISTS study evolution and function of mammals in the wild and within controlled environments. Mammals are researched, analyzed and studied including mammals giving live births such as monkeys, apes, whales, elephants and those that lay eggs such as platypus.

MARINE BIOLOGISTS research functions, structure and the evolution of animals and organisms living in saltwater.

ORNITHOLOGISTS scientifically study birds. Many in this field have degrees in other fields such as ecology or biology.

Scientists with Similar Disciplines

BIOCHEMISTS – Research and study large and small organisms including animals.

MICROBIOLOGISTS – Research and study bacteria, algae and fungi including how they grow and reproduce.

PHYSIOLOGISTS – Research and study entire life cycles of animals and plants.

REFERENCES

Agricultural and Animal Scientists

"Agricultural and Food Scientists." *U.S. Bureau of Labor Statistics*. U.S. Bureau of Labor Statistics, n.d. Web. Nov. 2013.

"What Does An Animal Scientist Do." *WiseGEEK*. N.p., n.d. Web. June-July 2013.

Botanists

"Botanist Job Description." *Career as a Botanist, Salary, Employment - Definition and Nature of the Work, Education and Training Requirements, Getting the Job.* N.p., n.d. Web. Oct.-Nov. 2013.

"Botanist Salary | Salary.com." *Salary.com*. N.p., n.d. Web. Dec. 2013.

"Botany Society of America." *Botanical Society of America, Leading Scientists and Educators since 1893.* N.p., n.d. Web. June-July 2012.

"Job Descriptions and Careers, Career and Job Opportunities." *Botany Careers*. N.p., n.d. Web. June-July 2012.

What can I do with a major in Botany? - November 2012 "What Can I Do with a Major in Botany?" *What Can I Do with a Major in Botany?* N.p., n.d. Web. June-July 2013.

Conservation Scientists

"Conservation Scientists." *Conservation Scientist Job Description*. N.p., n.d. Web. Nov. 2013.

"Conservation Scientists and Foresters." *Conservation Scientists and Foresters*. N.p., n.d. Web. Dec. 2013.

"Conservation Scientists and Foresters." *U.S. Bureau of Labor Statistics*. U.S. Bureau of Labor Statistics, n.d. Web. Oct. 2013.

"How Many People Are Currently Employed as a Conservation Scientists and Foresters in the United States?" *Career Questions RSS*. N.p., n.d. Web. Dec. 2013.

Environment Scientists

"Environmental Scientists and Specialists." *U.S. Bureau of Labor Statistics*. U.S. Bureau of Labor Statistics, n.d. Web. Nov.. 2013.

"How Many People Are Currently Employed as a Conservation Scientists and Foresters in the United States?" *Career Questions RSS*. N.p., n.d. Web. April. 2013.

"Hydrologists." *U.S. Bureau of Labor Statistics*. U.S. Bureau of Labor Statistics, n.d. Web. Dec. 2013.

Food Scientists

"Agricultural and Food Scientists." *U.S. Bureau of Labor Statistics*. U.S. Bureau of Labor Statistics, n.d. Web. Nov. 2013.

Foresters

"Conservation Scientists and Foresters." *Conservation Scientists and Foresters*. N.p., n.d. Web. Dec. 2013.

"Conservation Scientists and Foresters." *U.S. Bureau of Labor Statistics*. U.S. Bureau of Labor Statistics, n.d. Web. Oct. 2013.

"How Many People Are Currently Employed as a Conservation Scientists and Foresters in the United States?" *Career Questions RSS*. N.p., n.d. Web. June. 2013.

Veterinarians

"Job Markets for Uncommon Veterinarians Look Good." *Elliott Garber*. N.p., n.d. Web. Dec. 2013

Nature Jobs. *Nature Publishing Group*. n.d. Web. Dec. 2013.

"Veterinarians." *U.S. Bureau of Labor Statistics*. U.S. Bureau of Labor Statistics, n.d. Web. Mar. 2012.

Wildlife Biologists

"Zoologist." *About Bioscience*. N.p., n.d. Web. Dec. 2013.

"Zoologists and Wildlife Biologists." *U.S. Bureau of Labor Statistics*. U.S. Bureau of Labor Statistics, n.d. Web. Oct. 2013.

"Zoologists and Wildlife Biologists." *Zoologists and Wildlife Biologists*. N.p., n.d. Web. Dec. 2013.
"Zoologist Job Description." *Career as a Zoologist, Salary, Employment - Definition and Nature of the Work, Education and Training Requirements, Getting the Job*. N.p., n.d. Web. Oct.-Nov. 2013.

Zoologists

"Zoologist." *About Bioscience*. N.p., n.d. Web. Dec. 2013.

"Zoologists and Wildlife Biologists." *U.S. Bureau of Labor Statistics*. U.S. Bureau of Labor Statistics, n.d. Web. Oct. 2013.

"Zoologists and Wildlife Biologists." *Zoologists and Wildlife Biologists*. N.p., n.d. Web. Dec. 2013.

"Zoologist Job Description." *Career as a Zoologist, Salary, Employment - Definition and Nature of the Work, Education and Training Requirements, Getting the Job*. N.p., n.d. Web. Oct.-Nov. 2013.

Information and Resources

American Registry of Professional Animal Scientists (ARPAS)
1800 South Oak Street, Suite 100
Champaign, Illinois 61820-6974
http://www.arpas.org/

American Society of Animal Science (ASAS)
Post Office Box 7410,
Champaign, Illinois 61826-7410
http://www.asas.org

American Veterinary Medical Association (AVMA)
1931 North Meacham Road, Suite 100
Schaumburg, Illinois 60173
http://www.avma.org

Association of American Veterinary Medical Colleges (AAVMC)
1101 Vermont Avenue N.W., Suite 301
Washington, DC 20005
http://www.aavmc.org

The Botanical Society of America (BSA)
Post Office Box 299
St. Louis, Missouri 63166-0299
http://www.botany.org/

National Board of Veterinary Medical Examiners (NBVME)
Post Office Box 1356
Bismarck, North Dakota 58502
http://www.nbvme.org/

Information and Resources
(continued)

National Institutes of Health (NIH)
9000 Rockville Pike
Bethesda, Maryland 20892
http://nih.gov/

The National Science Foundation (NSF)
4201 Wilson Boulevard
Arlington, Virginia 22230
http://www.nsf.gov/

Society of American Foresters (SAF)
5400 Grosvenor Lane
Bethesda, Maryland 20814-2198
https://www.safnet.org/

Society for Range Management (SRM)
6901 South Pierce Street, Suite 225
Littleton, Colorado 80128
http://www.rangelands.org/

Soil Science Society of America (SSSA)
5585 Guilford Road
Madison, Wisconsin 53711-5801
https://www.soils.org/

Information and Resources

(continued)

United States Department of Agriculture (USDA)

Food Safety and Inspection Service
1400 Independence Avenue, S.W.
Washington, DC 20250-3700
http://www.fsis.usda.gov

United States Fish and Wildlife Service (USFWS)

http://www.fws.gov/

www.ingramcontent.com/pod-product-compliance
Lightning Source LLC
Chambersburg PA
CBHW022104170526
45157CB00004B/1481